THE POETRY OF LANTHANUM

The Poetry of Lanthanum

Walter the Educator

SKB

Silent King Books a WhichHead Imprint

Copyright © 2023 by Walter the Educator

All rights reserved. No part of this book may be reproduced in any manner whatsoever without written permission except in the case of brief quotations embodied in critical articles and reviews.

First Printing, 2023

Disclaimer
This book is a literary work; poems are not about specific persons, locations, situations, and/or circumstances unless mentioned in a historical context. This book is for entertainment and informational purposes only. The author and publisher offer this information without warranties expressed or implied. No matter the grounds, neither the author nor the publisher will be accountable for any losses, injuries, or other damages caused by the reader's use of this book. The use of this book acknowledges an understanding and acceptance of this disclaimer.

"Earning a degree in chemistry changed my life!"
- Walter the Educator

dedicated to all the chemistry lovers, like myself, across the world

CONTENTS

Dedication v

Why I Created This Book? 1

One - Oh, Lanthanum 2

Two - Night To Day 4

Three - Testament To The Wonders 6

Four - Fuel Creation 8

Five - Atomic Dance 10

Six - Plain Sight 12

Seven - A Spark 14

Eight - Element Of Grace 16

Nine - Jewel Beyond Compare 18

Ten - Ingenuity 20

Eleven - Element Precious 22

Twelve - Symbol Of Progress 24

Thirteen - Waiting To Be Told	. . . .	26
Fourteen - Elegance And Versatility	. . . . .	28
Fifteen - Intricate Prowess	. . . .	30
Sixteen - We Find Inspiration	. . . .	32
Seventeen - Humankind	. . . .	34
Eighteen - Endless Worth	. . . .	36
Nineteen - A Gem	. . . .	38
Twenty - Unique Properties	. . . .	40
Twenty-One - Remarkable	. . . .	42
Twenty-Two - Lanthanum, A Wonder	. . . . .	44
Twenty-Three - Brilliance And Might	. . . . .	46
Twenty-Four - Cosmic Space	. . . .	48
Twenty-Five - Shining Bright	. . . .	50
Twenty-Six - One And All	. . . .	52
Twenty-Seven - Concealed	. . . .	54
Twenty-Eight - Stars In The Night	. . . . . .	56
Twenty-Nine - Exploration's Call	. . . .	58
Thirty - Work Of Art	. . . .	60
Thirty-One - Relentless Dedication	. . . . .	62
Thirty-Two - Stand Tall	. . . .	64

Thirty-Three - Progress And Change 66

Thirty-Four - Steadfast And Pure 68

Thirty-Five - Endless Romance 70

About The Author 72

WHY I CREATED THIS BOOK?

Creating a poetry book about the chemical element Lanthanum was a unique and intriguing endeavor. Lanthanum, with its atomic number 57, is a rare earth element that possesses fascinating properties. By delving into the characteristics, history, and applications of Lanthanum, I can weave a narrative that explores themes of rarity, discovery, and the interconnectedness of science and art. Poetry has the power to make scientific concepts more accessible, engaging, and relatable, allowing readers to appreciate the beauty and wonder of the natural world in a different light. By creating this poetry book about Lanthanum, I can combine the realms of science and art, sparking curiosity and appreciation for both disciplines.

ONE

OH, LANTHANUM

In the realm of elements, a gem does shine,
Lanthanum, noble and truly divine.
A treasure of nature, a gemstone so rare,
In the depths of the earth, it lies with care.

With atomic number fifty-seven it stands,
A symbol of strength, in alchemical hands.
Soft and silvery, it gleams in the light,
Lanthanum, a marvel, both day and night.

Its name derived from the Greek word "lanthanein,"
To hide or escape, like secrets unseen.
For in its presence, wonders unfold,
A tale of marvels, yet to be told.

In ancient times, its power was unknown,
But now we uncover its secrets, alone.
Used in catalysts, ceramics, and lights,
Lanthanum, a helper, bringing delight.

From televisions to camera lenses,
It enhances our world with endless advances.
A conductor of electricity, strong and pure,
Lanthanum, a hero, forever endure.

In the dreams of scientists, it takes flight,
Unleashing potential with all of its might.
So let us celebrate this element grand,
Lanthanum, a jewel in nature's hand.

Oh, Lanthanum, you sparkle and gleam,
A symbol of progress, a scientific dream.
In the periodic table, you hold your place,
Forever enchanting, with grace and embrace.

TWO

NIGHT TO DAY

In the depths of the periodic table's frame,
A hidden gem, Lanthanum, gains its fame.
Element fifty-seven, its atomic decree,
A story untold, a mystery to see.

Silvery-white, its hue so serene,
Lanthanum, a realm where wonders convene.
With electrons dancing, in orbits they dance,
Creating a symphony of scientific romance.

Found in minerals, Earth's secret stowaway,
Lanthanum emerges, an element at play.
Ceramics and lenses, its talents employed,
Aiding progress, its virtues employed.

But beyond the lab, where dreams take flight,
Lanthanum's essence casts a radiant light.

A symbol of resilience, a spirit so pure,
It inspires the world, of that we can be sure.

From ancient lore to modern age,
Lanthanum's presence is a glorious stage.
A conductor of energy, a force so strong,
Igniting innovation, where dreams belong.

Oh, Lanthanum, your secrets untold,
In laboratories and stories of old.
A beacon of hope, a catalyst of change,
Lanthanum, in you, our futures arrange.

So let us embrace this element divine,
Lanthanum, a treasure, so rare and fine.
A symbol of progress, in every way,
Forever shining bright, from night to day.

THREE

TESTAMENT TO THE WONDERS

In the realm of elements, a mystic force does reside,
Lanthanum, the enigmatic, where secrets do hide.
Atomic number fifty-seven, a symbol of grace,
With a silvery gleam, it holds an ethereal embrace.

Beneath the Earth's surface, in minerals it dwells,
Lanthanum reveals its wonders, like hidden spells.
A catalyst for progress, a conductor of light,
It dances with electrons, a captivating sight.

From ancient alchemy to modern-day dreams,
Lanthanum's allure, like moonlit beams.
In laboratories, its potential unfurls,
Unleashing the mysteries of rare earth pearls.

Ceramics and lenses, it lends a helping hand,
Enhancing our world, like an artist's wand.

From televisions to hybrid car batteries,
Lanthanum's applications span vast territories.
   Yet, beyond its utility, it whispers a tale,
Of resilience and strength that will never fail.
A symbol of transformation, of inner alchemy,
Lanthanum, a reminder of boundless possibility.
   Oh, Lanthanum, you captivate and inspire,
Fueling innovation, setting our spirits on fire.
From the depths of the periodic table you arise,
Unveiling the magic hidden behind scientific ties.
   So let us celebrate this element profound,
Lanthanum, a jewel in nature's crown.
In its luminescence, we find hope and delight,
A testament to the wonders of nature's cosmic flight.

FOUR

FUEL CREATION

In the realm of elements, a gem so rare,
Lanthanum, a marvel beyond compare.
Silvery and soft, it graces the stage,
Unveiling secrets with wisdom and sage.

From ancient lands to laboratories anew,
Lanthanum's presence, a wonder to pursue.
Catalysts and lasers, its power unleashed,
Aiding progress, as boundaries are breached.

In the heart of technology, it finds its place,
Lanthanum, guardian of a digital space.
LED screens and electric cars, it empowers,
Guiding us forward, through innovation's towers.

But beyond the confines of science and steel,
Lanthanum's essence holds a deeper appeal.

A symbol of balance, of harmony and grace,
It reminds us to seek beauty in every space.
    Oh, Lanthanum, your elegance shines bright,
A beacon of hope in the darkest of night.
In the periodic table, you claim your throne,
A testament to nature's secrets, yet to be known.
    So let us celebrate this element divine,
Lanthanum, a treasure that forever will shine.
In its atomic dance, we find inspiration,
A reminder of the wonders that fuel creation.

FIVE

ATOMIC DANCE

In the realm of elements, a gem so rare,
Lanthanum, with mysteries beyond compare.
From the depths of Earth, it gracefully emerges,
Igniting curiosity, as science converges.

A conductor of energy, it sparks the flame,
Lanthanum, weaving wonders in its name.
In catalysts and alloys, its power is seen,
An alchemist's dream, a potent machine.

But beyond its utility, a story unfolds,
Lanthanum, a tale of secrets untold.
A symbol of resilience, it stands tall,
A silent guardian through it all.

Oh, Lanthanum, your presence revered,
A cosmic dancer that perseveres.

In laboratories, its essence explored,
Unleashing knowledge, like an open door.
　So let us honor this element rare,
Lanthanum, a marvel beyond compare.
In its atomic dance, we find inspiration,
A testament to nature's grand creation.

SIX

PLAIN SIGHT

In the realm of elements, a luminary appears,
Lanthanum, a treasure that captivates and endears.
With atomic number fifty-seven it resides,
A beacon of potential, where innovation resides.

Its name derived from the Greek word for hidden,
Lanthanum, a mystery waiting to be ridden.
Symbol La, it stands proud on the periodic chart,
Unveiling its wonders, igniting the spark.

From hybrid cars to wind turbines that spin,
Lanthanum, a catalyst for a world that's akin.
A conductor of electricity, a force so serene,
Empowering progress, like a visionary dream.

Oh, Lanthanum, your secrets we seek,
In laboratories and minds, fascination peaks.

Unveiling the marvels of rare earth's embrace,
Lanthanum, a testament to nature's grace.
  So let us celebrate this element divine,
Lanthanum, a jewel in science's shrine.
In its gentle radiance, we find hope and might,
A reminder of the wonders hidden in plain sight.

SEVEN

A SPARK

In the realm of elements, a star does shine,
Lanthanum, a treasure, both rare and fine.
With atomic grace, it dances in the core,
Whispering secrets of ancient cosmic lore.

A beacon of light, it illuminates the way,
Lanthanum, guiding us to a brighter day.
In cathode rays and lenses, its magic unfolds,
Revealing visions, stories yet untold.

Oh, Lanthanum, your essence, so pure,
A symbol of resilience that will endure.
From laboratories to technological advance,
You lead us forward in a cosmic dance.

In the depths of the Earth, your beauty resides,
A testament to nature's artistry, it presides.

With each atomic bond, a symphony plays,
Lanthanum, orchestrating nature's praise.
   So let us celebrate this element divine,
Lanthanum, a spark in the grand design.
In its luminescence, we find hope and grace,
A reminder of the wonders in every place.

EIGHT

ELEMENT OF GRACE

Lanthanum, oh element of grace,
In the realm of science, you find your place.
With atomic number fifty-seven you reside,
A gem of the periodic table, never to hide.

From the heart of Earth, you are unearthed,
A secret treasure, silently immersed.
Your presence, subtle yet strong,
Ignites imaginations, inspires a song.

In the alloys you form, strength you bestow,
Enhancing structures, helping them grow.
A catalyst for progress, you lead the way,
Innovations blooming, come what may.

Oh, Lanthanum, your luminescence shines bright,
A beacon of hope, a guiding light.
In the world of technology, you find your role,
From LED screens to rechargeable souls.

So let us celebrate your atomic dance,
Lanthanum, symbol of elegance and chance.
In your quiet beauty, a story unfolds,
Of a wondrous element, forever untold.

NINE

JEWEL BEYOND COMPARE

Lanthanum, an element of rare delight,
A symphony of electrons, shining so bright.
In the depths of the Earth, you patiently wait,
To reveal your secrets, to illuminate.

Your magnetic allure, a captivating spell,
Drawing us closer, like a tale to tell.
In lab coats and goggles, we seek your grace,
Unveiling mysteries, exploring your space.

Oh, Lanthanum, your atomic dance,
A rhythm that enchants, a cosmic romance.
In alloys and ceramics, your strength is found,
A building block, both sturdy and profound.

As a conductor of light, you paint the night,
With neon hues, a celestial sight.

In the hands of artists, you find your voice,
A muse for creation, a reason to rejoice.
   So let us celebrate this element rare,
Lanthanum, a jewel beyond compare.
In its quiet elegance, we find inspiration,
A symbol of wonder, a catalyst for creation.

TEN

INGENUITY

Lanthanum, a whisper in the air,
A siren song that weaves with care.
In the depths of the Earth, your story lies,
A tale of resilience, hidden from prying eyes.

In glowing embers, your presence gleams,
A catalyst of dreams, breaking through seams.
With each atomic step, you light the way,
Guiding scientists to discoveries, day by day.

Oh, Lanthanum, your magnetic charm,
Drawing us closer, like a lover's arm.
In magnets and lasers, your power resides,
Unleashing energy, where innovation abides.

From green phosphors to camera lenses clear,
You shape our world, without a trace of fear.

A guardian of progress, in elements you blend,
Lanthanum, a catalyst that knows no end.
   So let us raise a toast to this element divine,
Lanthanum, a treasure that forever shines.
In its silent brilliance, we find hope and grace,
A symbol of ingenuity, lighting up space.

ELEVEN

ELEMENT PRECIOUS

Lanthanum, oh element of grace,
In the depths of Earth, you find your place.
A hidden gem, rare and pristine,
Unveiling wonders, yet to be seen.
 In the laboratory's hallowed halls,
Scientists marvel at your atomic thralls.
With every reaction, you astound,
A symphony of elements, profound.
 Oh, Lanthanum, your presence sublime,
A catalyst for progress, throughout time.
From glassmaking to hybrid cars,
You play a vital role, reaching for the stars.
 In the realm of phosphors, you shine so bright,
Illuminating screens, casting a vibrant light.

A luminescent beauty, captivating our gaze,
Lanthanum, a marvel in this cosmic maze.
    So let us appreciate your rare allure,
Lanthanum, a symbol of dreams secure.
In your atomic dance, a tale unfolds,
Of an element precious, a story yet untold.

TWELVE

SYMBOL OF PROGRESS

Lanthanum, an element of grace,
In the vast universe, you find your place.
With atomic number fifty-seven,
Your presence in nature is a blessing from heaven.

A rare earth metal, with a lustrous sheen,
You captivate scientists, a visionary dream.
In your ionic state, you form compounds strong,
Aiding in technologies, where they belong.

Oh, Lanthanum, your magnetic might,
Draws us closer, like stars in the night.
From powerful magnets to MRI scans,
You assist medical breakthroughs, in healing hands.

In the world of green chemistry, you excel,
Reducing waste, protecting the environment's spell.

A catalyst for change, you inspire,
Pushing boundaries higher and higher.
 So let us honor your atomic art,
Lanthanum, a masterpiece from the heart.
In your presence, innovation takes flight,
A symbol of progress, shining ever so bright.

THIRTEEN

WAITING TO BE TOLD

In the realm of the periodic table, you reside,
Lanthanum, a treasure, in elements' stride.
With atomic grace and a silvery gleam,
You illuminate our world, like a radiant beam.

Oh, Lanthanum, your secrets unfold,
A story untold, yet to be told.
In the alloys you form, strength you impart,
Forging bridges of progress, connecting every heart.

From the screens that flicker, vibrant and bright,
To the lenses capturing nature's pure light,
You shape our vision, expanding our sight,
Lanthanum, a catalyst, igniting our insight.

In the depths of Earth, where your essence lies,
Miners toil, seeking your precious ties.

With dedication and sweat, they unearth,
The hidden gem that's filled with worth.
 So let us celebrate your atomic embrace,
Lanthanum, a symbol of ingenuity and grace.
In laboratories, your wonders unfold,
A testament to the stories, waiting to be told.

FOURTEEN

ELEGANCE AND VERSATILITY

Lanthanum, an element of grace,
In the periodic table, you find your place.
With atomic number fifty-seven,
You light up our world like a celestial beacon.
In the realm of cerium and praseodymium,
You stand tall, shining with your own rhythm.
A soft, silvery metal, rare and refined,
You captivate our minds, leaving us entwined.
Your magnetic properties, a wonder to behold,
In motors and generators, your power unfolds.
From hybrid cars to wind turbines,
You propel us forward, where progress aligns.
Oh, Lanthanum, your radiance inspires,
In phosphors and lasers, your brilliance transpires.

In the glow of televisions and fluorescent lights,
You paint the world with hues, day and night.
   As a catalyst of change, you spark innovation,
In chemical reactions, a source of transformation.
Lanthanum, a symbol of endless possibilities,
A tribute to the wonders of scientific discoveries.
   So let us cherish your atomic symphony,
Lanthanum, a marvel of our human journey.
In your elegance and versatility, we find,
A testament to the wonders of the elemental design.

FIFTEEN

INTRICATE PROWESS

Lanthanum, an element rare and true,
Your presence shines, a celestial hue.
In the depths of Earth, where secrets lie,
You await discovery, beneath the sky.

Oh, Lanthanum, a conductor of light,
In fiber optics, you guide our sight.
Through invisible threads, signals dance,
Connecting the world in a cosmic trance.

In rechargeable batteries, you find your place,
Empowering devices, with steady embrace.
From smartphones to laptops, you fuel our needs,
A silent force, where progress succeeds.

With your magnetic charm, you enchant,
In motors and generators, you grant

The energy to move, the power to thrive,
Lanthanum, the spark that keeps us alive.

    So let us celebrate your atomic grace,
Lanthanum, a marvel in this vast space.
In your silent presence, we find solace,
A reminder of nature's intricate prowess.

SIXTEEN

WE FIND INSPIRATION

Oh, Lanthanum, rare and bright,
In the realm of elements, a guiding light.
With atomic number 57, you stand,
A symbol of wonder, crafted by nature's hand.

In the world of steel, you lend your strength,
An alloy companion, going to great lengths.
From skyscrapers to bridges, you provide,
The resilience and support, standing side by side.

Through the lenses of cameras, your vision is clear,
Capturing moments, memories held dear.
In glassmaking, your influence reigns,
Creating beauty, where artistry remains.

A catalyst you are, in chemical reactions,
Driving progress with your atomic transactions.

From petroleum refining to polymerization,
Lanthanum, a key to innovation's ignition.
 So let us raise our voices, in awe and acclaim,
For Lanthanum, a jewel in science's domain.
In your versatile nature, we find inspiration,
A testament to the universe's creation.

SEVENTEEN

HUMANKIND

Lanthanum, oh element of wonder,
In the realm of the periodic thunder.
Your atomic number fifty-seven,
A symbol of grace, a gift from heaven.

In the depths of Earth, where you reside,
Mineral veins where secrets hide.
Mined with care, your treasure sought,
A precious gem, with value unbought.

With magnetic properties, you hold sway,
Guiding compasses, pointing the way.
In magnets strong, you play a part,
Drawing us closer, heart to heart.

In the lab, your secrets unfold,
Catalyst of reactions, stories untold.

Chemical transformations, you facilitate,
Unleashing potential, a power innate.

Lanthanum, in phosphors you shine,
Radiant hues, a spectacle divine.
From neon lights to vibrant screens,
You illuminate our technological dreams.

So let us celebrate your atomic grace,
Lanthanum, a marvel in this cosmic space.
In your unique properties, we find,
A testament to the wonders of humankind.

EIGHTEEN

ENDLESS WORTH

In the realm of elements, you stand tall,
Lanthanum, a treasure, admired by all.
With atomic allure, you capture our gaze,
Unveiling secrets in a mystical haze.

In the depths of the Earth, where you reside,
Miners toil, seeking your hidden stride.
With unwavering dedication and might,
They bring you forth, into the light.

In metallurgy, you lend your strength,
Alloying with metals, going to great lengths.
Enhancing properties, forging a bond,
Creating materials, both resilient and fond.

Lanthanum, a conductor of electricity's reign,
In wires and circuits, you help us sustain,

The flow of energy, powering our world,
A symphony of electrons, gracefully twirled.
   So let us honor your atomic grace,
Lanthanum, a marvel in this cosmic place.
In your versatility, we find endless worth,
A testament to the wonders of the Earth.

NINETEEN

A GEM

Lanthanum, a beacon of rare delight,
In the realm of elements, you shine so bright.
With atomic number fifty-seven you stand,
A symbol of wonder, crafted by nature's hand.

In the land of magnets, you hold your sway,
Aligning domains, in a magnetic ballet.
From motors to MRI machines,
You power the world with magnetic dreams.

In the realm of glass, you find your place,
Enhancing clarity with elegant grace.
Optical lenses crafted with your touch,
Revealing beauty that means so much.

Lanthanum, a catalyst in chemistry's dance,
Igniting reactions with a subtle glance.
From petroleum refining to organic synthesis,
You facilitate progress with finesse.

So let us celebrate your atomic reign,
Lanthanum, a gem that's never plain.
In your versatile nature, we find inspiration,
A testament to the marvels of creation.

TWENTY

UNIQUE PROPERTIES

In the depths of the Earth, you reside,
Lanthanum, an element we cannot hide.
A rare gem, you shimmer and gleam,
A treasure of the periodic scheme.

In the world of lighting, you play a part,
Phosphors ablaze, a radiant art.
Fluorescent tubes, a vibrant glow,
Guiding us through darkness, as we go.

In the realm of chemistry, you bring change,
A catalyst, a force to rearrange.
Reactions unfold under your command,
Unlocking new compounds, vast and grand.

Lanthanum, a conductor of heat,
In high-temperature alloys, your presence is neat.

With strength and durability, you endure,
Forging materials that remain pure.
 So let us marvel at your atomic grace,
Lanthanum, a symbol of scientific embrace.
In your unique properties, we find,
A testament to the wonders of humankind.

TWENTY-ONE

REMARKABLE

Lanthanum, a shining star of the periodic table,
A rare gem, with allure that's undeniable.
In the deep earth, your treasures abide,
Mined with care, where secrets reside.

In the world of steel, you lend your might,
Strengthening structures, sturdy and tight.
With resilience and toughness, a steadfast aid,
Building bridges and towers, where dreams are made.

In the realm of medicine, you hold a key,
In X-ray machines, revealing what's unseen.
Diagnosing ailments, with precision and care,
Guiding doctors on paths to healing repair.

Lanthanum, a conductor of sound,
In speakers and headphones, melodies abound.
Transmitting vibrations, harmonies arise,
Filling the air, sparking joy in our eyes.

So let us celebrate your atomic grace,
Lanthanum, an element of remarkable embrace.
In your versatile nature, we find inspiration,
A testament to the wonders of creation.

TWENTY-TWO

LANTHANUM, A WONDER

Lanthanum, a hidden gem of the Earth's chest,
A celestial dancer, the cosmos' bequest.
With electrons swirling in orbits grand,
You hold secrets of the universe in your hand.

In the realm of lasers, you come alive,
Harnessing light, a spectacle to thrive.
From barcode scanners to laser shows,
You mesmerize with your radiant glows.

In the field of energy, you play a role,
Powering batteries, charging our soul.
From hybrid cars to renewable might,
You fuel our future, shining so bright.

Lanthanum, a guardian of the flame,
In lantern mantles, you earn your fame.

Casting a glow, guiding paths afar,
A beacon of warmth, beneath the stars.
 So let us embrace your atomic grace,
Lanthanum, a wonder in this cosmic space.
In your versatility, we find endless worth,
A testament to the wonders of planet Earth.

TWENTY-THREE

BRILLIANCE AND MIGHT

Lanthanum, a gem of the periodic table,
In the realm of elements, you're truly stable.
With atomic number fifty-seven, you reside,
A quiet presence, a force we cannot hide.

In the depths of Earth, you lie concealed,
Miners seek you, with a determined zeal.
Your ores, like treasures, they unearth,
Revealing your nature, a substance of great worth.

Lanthanum, a conductor of electricity's might,
In wires and circuits, you shine so bright.
With conductivity, you pave the way,
Powering devices, day after day.

In the lab, your mysteries unfold,
Catalyzing reactions, stories yet untold.

Scientists marvel at your chemical dance,
Creating compounds, advancing the expanse.
   So let us celebrate your atomic grace,
Lanthanum, a marvel of this cosmic space.
In your unique properties, we find delight,
A testament to nature's brilliance and might.

TWENTY-FOUR

COSMIC SPACE

In the realm of elements, you stand tall,
Lanthanum, a name that echoes with a call.
A rare earth gem, with a lustrous gleam,
A quiet presence, yet a vibrant beam.

In the world of alloys, you find your place,
Strengthening metals with your embrace.
From aerospace to automotive design,
You lend resilience, making them fine.

Lanthanum, a conductor of heat,
In high-temperature applications, you compete.
With stability and endurance, you endure,
Forging materials that remain pure.

In the land of glass, you weave your spell,
Enhancing clarity, a secret you tell.

Optical lenses crafted with your art,
Revealing beauty, stirring the heart.
   So let us honor your atomic grace,
Lanthanum, a marvel in this cosmic space.
In your versatility, we find inspiration,
A testament to the wonders of creation.

TWENTY-FIVE

SHINING BRIGHT

Lanthanum, a beacon from afar,
A luminescent celestial star.
With atomic number fifty-seven,
In the periodic table, you're heaven.
   In the world of magnets, you hold the sway,
Aligning fields, in a magnetic display.
From generators to MRI machines,
You power progress, fulfilling dreams.
   In chemistry's realm, you catalyze,
Driving reactions, a chemical guise.
From petroleum refining to polymer synthesis,
You facilitate transformations with finesse.
   Lanthanum, a conductor of electricity,
In wires and circuits, you flow with glee.

With conductivity, you light the path,
Powering innovation, with boundless aftermath.
    In the realm of glass, you leave your mark,
Improving optics, illuminating the dark.
With clarity and brilliance, you enhance,
Revealing the world, in a vibrant dance.
    So let us celebrate your atomic might,
Lanthanum, an element shining bright.
In your diverse properties, we find fascination,
A testament to nature's intricate creation.

TWENTY-SIX

ONE AND ALL

Lanthanum, an element of rare allure,
In the periodic table, you endure.
With an atomic number of fifty-seven,
Your presence in nature is a heavenly blessing.

In the world of magnets, you find your place,
As a component of neodymium's embrace.
Creating powerful magnets, strong and true,
With applications in technology, through and through.

Lanthanum, a conductor of heat,
In high-temperature alloys, you compete.
With strength and stability, you hold firm,
Forging materials that can withstand any squirm.

In the land of lighting, you shine so bright,
As phosphors in lamps, emitting colorful light.

From televisions to energy-saving bulbs,
You illuminate the world with vibrant pulsating pulse.
　So let us honor your atomic grace,
Lanthanum, a marvel in this cosmic space.
In your versatility, we find inspiration,
A testament to the wonders of creation.
　From medicine to industry, you make your mark,
Lanthanum, a catalyst igniting a spark.
In the vast universe of elements, you stand tall,
A symbol of science's triumph, for one and all.

TWENTY-SEVEN

CONCEALED

In the depths of Earth, you lie concealed,
Lanthanum, a treasure yet to be revealed.
A rare element, with atomic might,
You captivate our senses, like stars in the night.

In the realm of magnets, you hold sway,
With magnetic properties, you lead the way.
From powerful motors to magnetic storage,
You create a magnetic force, an endless voyage.

Lanthanum, a luminary in the world of glass,
Adding brilliance and clarity, a breathtaking mass.
In camera lenses and optical devices,
You bring clarity, revealing nature's priceless vices.

In the land of catalysts, you play a role,
Speeding reactions, an essential goal.

With your presence, chemical bonds transform,
Creating new compounds, a chemical storm.
 So let us celebrate your atomic grace,
Lanthanum, a wonder in this cosmic space.
In your versatility, we find inspiration,
A testament to the wonders of creation.

TWENTY-EIGHT

STARS IN THE NIGHT

Lanthanum, a jewel in the periodic array,
An element of wonder, in its unique display.
With atomic number fifty-seven, it claims its place,
A versatile element, worthy of embrace.

In the realm of electronics, you take the lead,
Conducting electricity with remarkable speed.
From smartphones to computers, you power their core,
Enabling connectivity, forevermore.

Lanthanum, a guardian of light,
In phosphors, you shimmer, gleaming so bright.
From televisions to fluorescent lamps,
You illuminate the world, dispelling the damp.

In the realm of chemistry, you catalyze,
Facilitating reactions, like a magician's disguise.

From organic synthesis to refining crude,
You accelerate transformations with gratitude.
   So let us celebrate your atomic grace,
Lanthanum, a marvel in this cosmic space.
In your diverse properties, we find fascination,
A testament to nature's intricate creation.
   Lanthanum, a symbol of ingenuity and might,
An element that shines, like stars in the night.
In the vast tapestry of elements, you stand tall,
A catalyst of progress, inspiring one and all.

TWENTY-NINE

EXPLORATION'S CALL

In the realm of elements, you hold your ground,
Lanthanum, a treasure yet to be fully found.
With atomic number fifty-seven, you reside,
A symbol of scientific progress, our guide.

Lanthanum, a conductor of electricity,
In circuits and wires, you display your ability.
With conductivity and strength, you empower,
Fueling innovation, hour after hour.

In the land of glass, you leave your mark,
Enhancing optical clarity, a vibrant spark.
From camera lenses to telescope glass,
You reveal the world, making it unsurpassed.

In the realm of chemistry, you shine bright,
Catalyzing reactions, a catalyst's delight.

From petroleum refining to polymerization,
You facilitate transformations, a source of fascination.
   So let us celebrate your atomic grace,
Lanthanum, a marvel in this cosmic space.
In your versatility, we find inspiration,
A tribute to nature's intricate creation.
   Lanthanum, a symbol of progress and dreams,
A beacon of hope, as science gleams.
In the vast universe of elements, you stand tall,
A testament to human curiosity, exploration's call.

THIRTY

WORK OF ART

In the realm of elements, you hold your might,
Lanthanum, shining with a celestial light.
With atomic number fifty-seven, you stand,
A gem among the elements, both rare and grand.
 In the field of metallurgy, you excel,
Strengthening alloys, like an alchemist's spell.
From aircraft frames to sturdy machines,
You lend strength, making them robust and keen.
 Lanthanum, a conductor of heat and electricity,
In circuits and wires, you flow with efficacy.
With conductivity, you keep the current alive,
Empowering technology to thrive and strive.
 In the world of glass, you weave your art,
Enhancing optics, creating a mesmerizing chart.

From camera lenses to telescopes afar,
You bring clarity, unveiling the universe's star.
   So let us celebrate your atomic grace,
Lanthanum, a treasure in this vast cosmic space.
In your versatility and rare presence,
We marvel at nature's intricate essence.
   Lanthanum, a symbol of ingenuity profound,
A testament to scientific wonders that astound.
In the symphony of elements, you play your part,
An inspiration to human knowledge, a true work of art.

THIRTY-ONE

RELENTLESS DEDICATION

Lanthanum, a gem among the elements' array,
A catalyst for progress in its own special way.
In the realm of magnets, you hold great might,
With magnetic properties, you shine so bright.

From powerful motors to magnetic resonance,
You bring forth innovation, a magnetic presence.
In the world of alloys, you lend your strength,
Creating materials that will go to any length.

Lanthanum, a luminary in the land of light,
You illuminate our lives, like stars in the night.
As phosphors in displays, you dazzle and gleam,
Enchanting our senses with a vibrant beam.

In the realm of chemistry, you play a role,
Aiding reactions, a catalyst for the soul.

From organic synthesis to fuel refinement,
With your atomic magic, you bring refinement.
   So let us celebrate your atomic grace,
Lanthanum, a marvel in this cosmic space.
In your versatility, we find inspiration,
A symbol of science's relentless dedication.

THIRTY-TWO

STAND TALL

Lanthanum, an element so rare and bright,
In the vast expanse of the periodic table's might.
With atomic number fifty-seven, you reside,
A symbol of wonder, impossible to hide.

In the world of magnets, you hold sway,
Your magnetic properties, a powerful display.
From motors to generators, you create a force,
A magnetic field that guides our course.

Lanthanum, a luminary in the realm of glass,
Adding brilliance and clarity, like a looking glass.
In camera lenses and prisms, you cast a spell,
Revealing the world in a way none can excel.

In the realm of catalysts, you play your part,
Speeding up reactions with scientific art.
From chemical transformations to industry's needs,
You facilitate progress with remarkable speed.

So let us honor your atomic grace,
Lanthanum, a marvel in this cosmic space.
In your versatility, we find inspiration,
A symbol of science's unwavering dedication.

Lanthanum, a beacon of progress and light,
An element that shines with all its might.
In the grand tapestry of elements, you stand tall,
A reminder of nature's wonders, captivating all.

THIRTY-THREE

PROGRESS AND CHANGE

Lanthanum, a hidden treasure of the Earth,
A marvel of the periodic table's birth.
With atomic number fifty-seven you reside,
In the rare earth metals, you take pride.

In magnetic fields, your presence is felt,
A conductor of electricity, like a cosmic belt.
From electric cars to wind turbines high,
You power the future with a sustainable sigh.

In the realm of medicine, you lend a hand,
As a contrast agent, you help doctors understand.
In imaging tests, you illuminate the way,
Guiding diagnoses with precision each day.

So let us celebrate your atomic grace,
Lanthanum, a marvel in this cosmic space.

In your versatility and scientific acclaim,
We admire your contributions, deserving of fame.
Lanthanum, a symbol of progress and change,
A catalyst for innovation, a range.
In the vast spectrum of elements, you stand tall,
A testament to human ingenuity, inspiring us all.

THIRTY-FOUR

STEADFAST AND PURE

Lush and luminous, you grace the Earth's stage,
Lanthanum, an element of wisdom and sage.
In the depths of the periodic table, you reside,
A rare gem, with secrets you confide.

In the realm of magnets, you possess a spell,
With magnetic properties, you weave a compelling tale.
From motors to generators, you bring power to life,
Magnetic fields swirling, buzzing with strife.

In the world of steel, you lend your might,
Strengthening structures, with a gleaming light.
From bridges to buildings, you fortify,
Ensuring resilience, reaching for the sky.

So let us celebrate your atomic grace,
Lanthanum, a gem in this cosmic space.

In your versatility, we find fascination,
A symbol of science's endless exploration.
   Lanthanum, a symbol of strength and allure,
A beacon of progress, steadfast and pure.
In the grand symphony of elements, you play a part,
A reminder of nature's wonders, captivating the heart.

THIRTY-FIVE

ENDLESS ROMANCE

Lanthanum, a luminary rare,
Shining brightly beyond compare.
In the realm of elements, you reside,
With atomic grace, you never hide.

In the world of glass, you take command,
Crafting beauty with a skilled hand.
From delicate lenses to prisms bold,
You transform light, a story untold.

With magnetic allure, you hold the key,
Guiding electrons, setting them free.
In motors and batteries, you play a role,
Powering machines, a vital soul.

So let us celebrate your atomic might,
Lanthanum, a beacon of light.

In your versatility, we find true worth,
A marvel of nature's extraordinary birth.
    Lanthanum, a symbol of curiosity's fire,
Igniting discoveries that never tire.
In the grand tapestry of elements, you dance,
A testament to science's endless romance.

ABOUT THE AUTHOR

Walter the Educator is one of the pseudonyms for Walter Anderson. Formally educated in Chemistry, Business, and Education, he is an educator, an author, a diverse entrepreneur, and he is the son of a disabled war veteran. "Walter the Educator" shares his time between educating and creating. He holds interests and owns several creative projects that entertain, enlighten, enhance, and educate, hoping to inspire and motivate you.

Follow, find new works, and stay up to date
with Walter the Educator™
at WaltertheEducator.com

www.ingramcontent.com/pod-product-compliance
Lightning Source LLC
LaVergne TN
LVHW051959060526
838201LV00059B/3736